Algebra Is for Children

Algebra Is for Children

JULIO CÉSAR MARTÍNEZ ROMERO

Library of Congress Control Number:		2017901877
ISBN:	Hardcover	978-1-5065-1881-7
	Softcover	978-1-5065-1880-0
	eBook	978-1-5065-1879-4

Print information available on the last page.

Rev. date: 06/02/2017

To order additional copies of this book, contact:
Palibrio
1663 Liberty Drive
Suite 200
Bloomington, IN 47403
Toll Free from the U.S.A 877.407.5847
Toll Free from Mexico 01.800.288.2243
Toll Free from Spain 900.866.949
From other International locations +1.812.671.9757
Fax: 01.812.355.1576
orders@palibrio.com
756216

Contents

I thank Daniela Carrera Millán and Natalia Pereyra Millán for their kind participation in this book. I also thank and acknowledge the work of Ismael Álvarez León, who kindly took and donated the cover photograph.

I

It was just another day. Mama Coty woke up early. Her little brother and her three daughters were still sleeping on their beds. Mama Coty looked at them, she smiled and went out to look for some fresh food for the day. When she arrived to a log that she used as a bridge to cross the stream every day, she looked around very carefully. You could never be too careful. Due to their small size, all sorts of danger threatened ocelot people. Birds of prey, like the hungry harpy eagles that soared high above in the sky, were always looking for a meal. An ocelot could be such a meal. Mama Coty had also occasionally caught the musky scent of a jaguar, an encounter she should avoid at all cost.

Mama Coty always crossed the stream on the same spot, not only because she knew the place by heart but also because alligators never ventured this high up the stream. She feared gators more than anything else in the world, even more than she feared the poachers who set snares and traps to catch ocelot children and then sold them as exotic pets.

Mama Coty always ran over the log as fast as she could. This day as she was running, she saw a huge shadow under the water moving towards her. A gigantic alligator raised its huge head and leaped to the log to grab her with its enormous jaws. Mama Coty jumped up as high as she could, trying to reach the branches of a tree leaning above the stream.

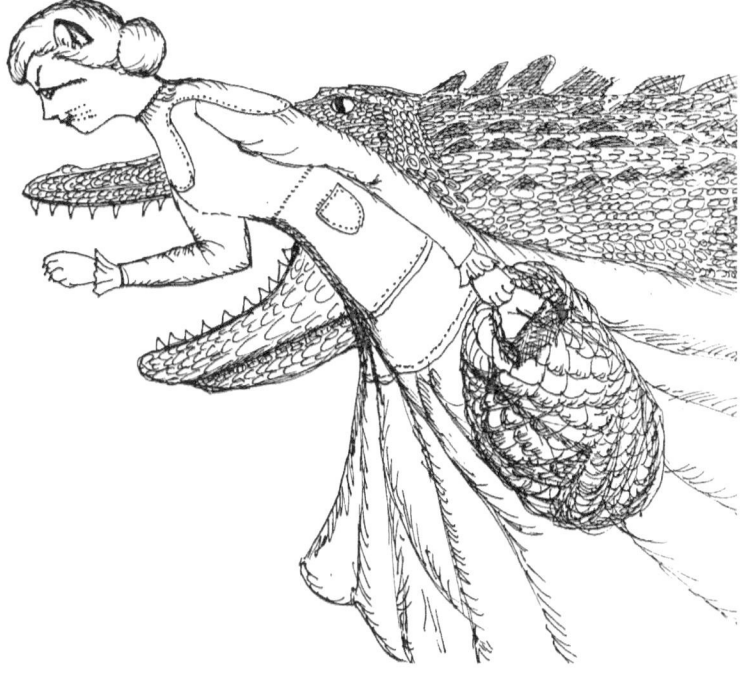

Fig.1

Tere was the first of the ocelot children to wake up. Esperanza and Clarita were still on their beds. Her young uncle Felix's eyes suddenly opened. He looked around.

- Coty has not come back yet. - Felix said.
- Where is Mom? - Clarita whispered.
- Why hasn't she come back? - Esperanza asked.

The four ocelot cubs wandered shyly out of their den and walked towards Mama Coty's hunting grounds. First, they advanced fearfully and as their confidence improved, they started walking decisively and briskly, except for Clarita, who was always a little frightened. They went up the bank of the river to the log where they knew their mother always crossed to the other side.

Fig. 2

Mama Coty was hanging from a branch, fearing that it would break at any moment or that she would grow tired, lose her grip and fall into the open jaws of the alligator. She heard Esperanza's voice.

- Mother, hold fast. We will rescue you.

The four children were already throwing stones inside the open mouth of the gator and were trying to poke its eyes with sticks. As the gator turned to the cubs, Mama Coty fell back on the log and hurried back to her children.

Fig. 3

The five of them ran as fast as they could and did not stop until they reached the safety of their den.

- In my whole life, I had never heard of a gator swimming this high up the stream. - Mama Coty said as soon as she could catch up her breath. - I cannot stand this. We will move away immediately. Right now! We will not stay a single moment more in this dangerous place.

Mama Coty was at her wits end, so frightened she was of the alligators.

The five ocelots traveled for two weeks. They crossed the forest, the jungle and the swamps. It took them twice as long to walk through the jungle as it had taken them to travel through the forest. To cross the swamps it took them twice as long as it took them to travel through the jungle.

How long did it take them to cross each region?

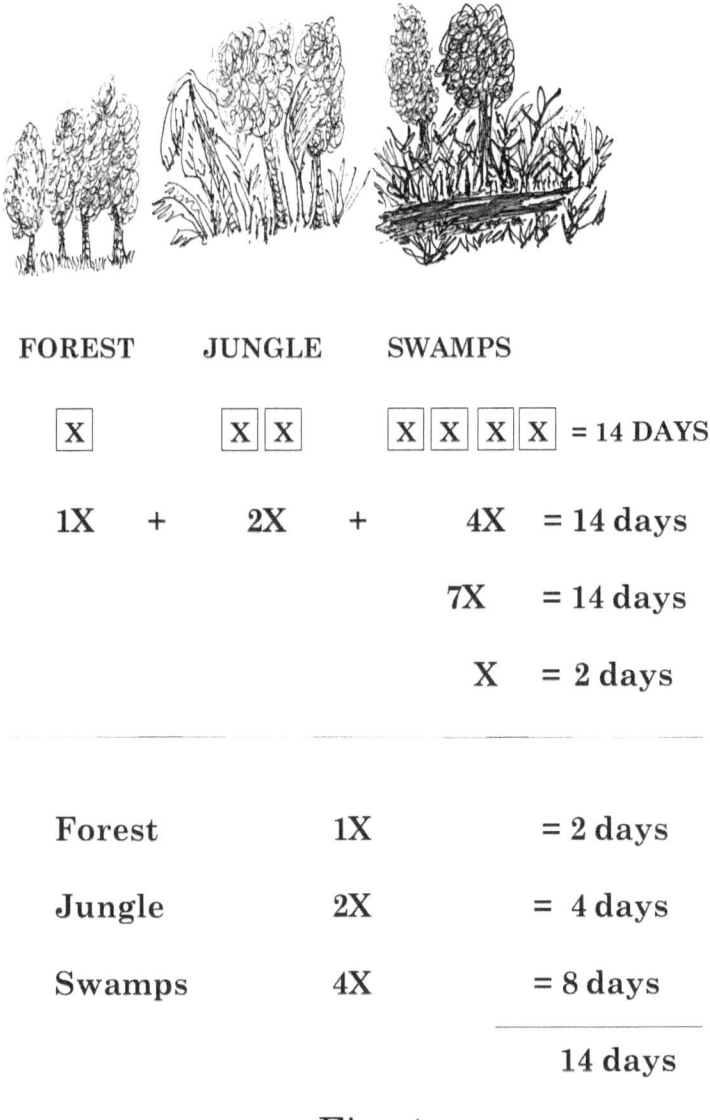

FOREST   JUNGLE   SWAMPS

| X | | X || X | | X || X || X || X | = 14 DAYS

1X   +   2X   +   4X   = 14 days

7X   = 14 days

X   = 2 days

Forest   1X   = 2 days

Jungle   2X   = 4 days

Swamps   4X   = 8 days

14 days

Fig. 4

After two days in the forest, four days in the jungle and eight days in the swamps, they arrived to a village. During their journey, they had been gathering eggs. Mama Coty had collected more eggs than anyone else, thrice as many as Felix. Felix had twice as many as Esperanza. Esperanza and Tere had the same amount and Clarita had two less than Tere. If Esperanza had 12 eggs, how many eggs did each of them gather? How many did they collect all together?

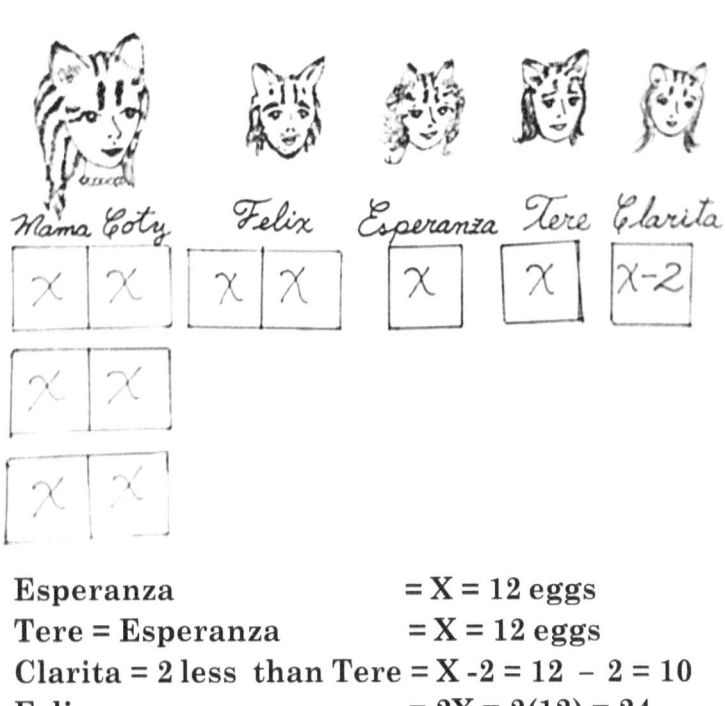

Esperanza	= X = 12 eggs
Tere = Esperanza	= X = 12 eggs
Clarita = 2 less than Tere	= X -2 = 12 − 2 = 10
Felix	= 2X = 2(12) = 24
Mama Coty	= 3(2X) = 3(24) = 72

Total	130

Fig. 5

Once in the city, the five ocelots took a room in an inn. Mama Coty fixed a 10 eggs omelet. After breakfast, they went to the market to inquire how much they would get if they sold the remaining eggs. They were told that they would get a roasted quail for each dozen eggs. How many roasted quails did they receive?

Mama Coty gave each ocelot the same number of quails. How many quails did each ocelot get?

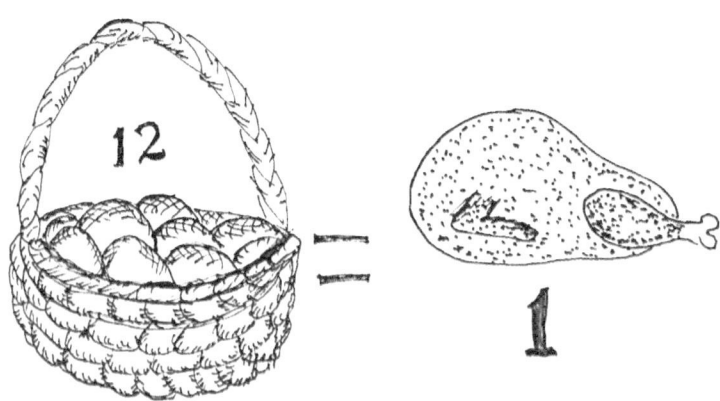

130 eggs – 10 eggs in the omelet= 120 eggs

120 eggs = 10(12) eggs

120 eggs = 10 dozen eggs

120 eggs = 10 roasted quails

10 roasted quails= 2 quails(5 ocelots)

10 roasted quails= 2 quails each

Fig. 6

Each ocelot ate one quail for dinner and saved the second one for the next stage of their trip. The following morning they woke up and started their journey to the mountains. Mama Coty had been told that there were magnificent meadows at the other side of the mountains, safe from men and alligators and full of quarry. Mama Coty set up her mind to go there to establish themselves.

They had walked for about an hour when Clarita realized that she had forgotten her remaining roasted quail in the hotel room. Without telling anyone where she was going, Clarita ran back to the inn.

At the door of the inn, Clarita paused for a while to catch her breath and walked in. There was a troop of squirrel monkeys at the hall. They all watched her with interest and exchanged weird comments that she could not understand. When Clarita came back from the room with her roasted quail, the squirrel monkeys started teasing her and hindered her exit. She did not know what to do and was very scared.

Fig. 7

Not long after Clarita had left the ocelot family, the rest of the party noticed her absence. Her mother, sisters and uncle looked for her and called her name hoping that she would answer. Sadly, there was no hint of her whereabouts. After pondering on the situation, the four of them traced back their steps and returned to the village. Nowhere could Clarita be found.

Fig. 8

They walked back to the inn and asked if anyone had seen the girl. They soon learned what had happened. A gang of squirrel monkeys had kidnapped her and had taken her to their headquarters in the middle of the jungle. A couple of weeks before, the Prince of the squirrel monkey kingdom had been taken as a pet by humans. In order to rescue him, humans had asked the squirrel monkeys for a new pet in exchange for their Prince. The squirrel monkeys hoped that Clarita would be a suitable substitute and that she would help them recover their Prince.

II

To get back to the jungle, Mama Coty and her family had to go back through the swamps. How long had it taken the five of them to cross the swamps on their way to the village?

Mama Coty feared that it would take them too long to find Clarita. She was also afraid that the squirrel monkeys might have closed their deal with the humans and had exchanged Clarita for their Prince. Therefore, Mama Coty asked a manatee to let them ride on its back in order to cross the swamps more quickly.

- Dear manatee, we need to cross the swamps as fast as possible so that I can rescue my daughter. Please let us ride on your back to the border of the jungle. - asked Mama Coty.

- I will gladly do as you request. Nevertheless, I have something to ask in return. My crown used to have ten pearls on it. One day some fish stole one pearl each and only three are left. If you bring me back my pearls, I will let you ride on my back through the streams to the border of the jungle.

X pearls stolen +   3 left in the crown= total

X   +   3   = 10

Fig. 9

Mama Coty went to look for the fish and explained to them why she needed their help.

- Dear fish, please give me back the pearls that you took from the manatee's crown so that it will let us ride on its back across the swamps as fast as possible to the border of the jungle so that I can rescue my daughter. - asked Mama Coty.

- We will gladly do as you request. Nevertheless, we have something to ask in return. The branches of a mango tree hang above the stream. The fruit used to fall in the stream, attracting insects that we ate. One day, three macaws took our mangoes and now we have no insects to eat. If you bring us back the mangoes that a single macaw took, there will be enough insects for us to eat. Then we will gladly give you the pearls that you need.

- How many mangoes did each macaw take?

- In all there were 18 mangoes, three fell in the forest and the others were equally distributed among the three macaws.

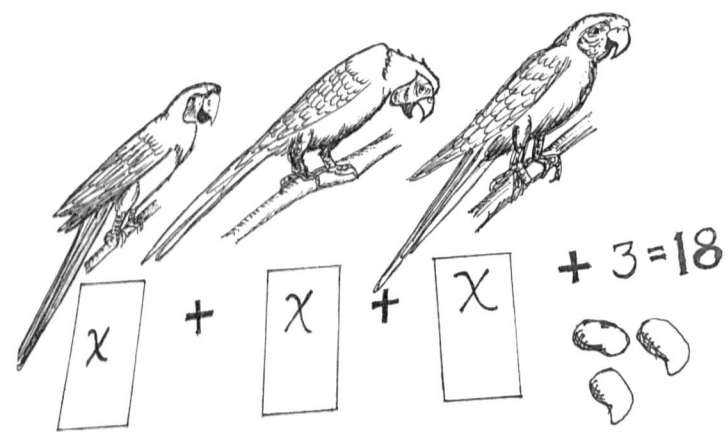

3 macaws with   3 mangoes fell
X mangoes each   in the forest =   total

$$3X \quad + \quad 3 \quad = \quad 18$$

Fig. 10

Mama Coty went to look for the macaws and explained to them why she needed their help.

- Dear macaws, please give me the mangoes that one of you have so that the fish will give me back the pearls that they took from the manatee's crown and then it will let us ride on its back across the swamps as fast as possible to the border of the jungle so that I can rescue my daughter. - asked Mama Coty.

- We will gladly do as you request. Nevertheless, we have something to ask in return. Please ask the raccoon to give us the shrimps that he catches in the stream on one day. - The macaws replied.

- How many shrimps are those?

- If you go to see him right now, he will be catching shrimps and he will inform you. - The macaws said.

Mama Coty and her family went to look for the raccoon. He was very busy.

- Instead of standing there just watching me, why don't you help me catch shrimps? - The raccoon said when he saw them.

Each of the ocelots caught a shrimp.

- How many shrimps do you catch each day? - Mama Coty asked him.

- In three days, I will have 34, including the 4 that you just caught.

3 days collecting+ shrimps		one shrimp collected by each ocelot	=	total
3X	+	4	=	34

Fig. 11

- Dear Mr. Raccoon, please give me the shrimps that you catch on a day. I will give them to the macaws. In turn, one of them will give me its mangoes. In exchange of the mangoes, the fish will give me back the pearls that they took from the manatee's crown and then it will let us ride on its back across the swamps as fast as possible to the border of the jungle so that I can rescue my daughter. - asked Mama Coty.

- I will gladly do as you request. In return, please ask one of the four squirrels to give me back the hazelnuts that it took away from me. - The raccoon replied.

- How many hazelnuts does each squirrel have?

- The four squirrels have 90 hazelnuts in all, including 10 in a basket. The rest of the hazelnuts were distributed equally among the four squirrels.

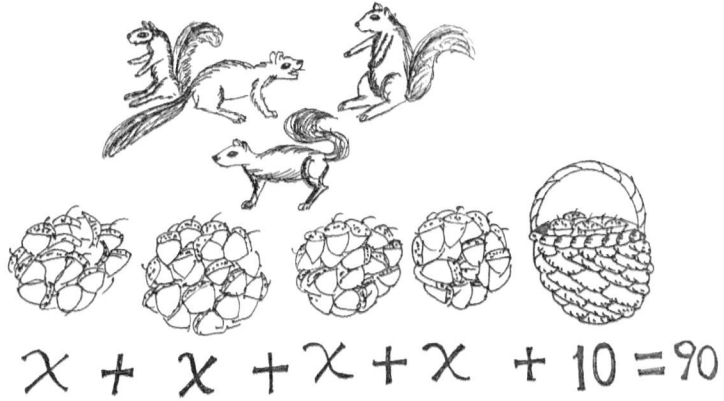

$$\chi + \chi + \chi + \chi + 10 = 90$$

4 squirrels each with X hazelnuts +	10 hazelnuts in a basket	= total
4X +	10	= 90

Fig. 12

Mama Coty and her family went to look for the squirrels.

When the ocelots found the squirrels, Mama Coty asked them.

- Will you please give me the hazelnuts that any of you have? I will give them to the raccoon. He will give me the shrimps that he catches on a day. I will give the shrimps to the macaws. In turn, one of them will give me its mangoes. In exchange of the mangoes, the fish will give me back the pearls that they took from the manatee's crown and then it will let us ride on its back across the swamps as fast as possible to the border of the jungle so that I can rescue my daughter. - asked Mama Coty.

-We will give you nothing. The hazelnuts are ours and we will not share them. - Was their answer.

The four ocelots jumped and each one of them caught a squirrel.

- We, ocelots, eat squirrels. If you don't give us the hazelnuts, we will eat you right away. - Mama Coty warned them.

Thus, the squirrels gave them the hazelnuts.

Fig. 13

As soon as the ocelots had the hazelnuts, Mama Coty gave them to the raccoon. He gave her the shrimps that he caught on a day. Mama Coty gave the shrimps to the macaws. In turn, one of them gave her its mangoes. In exchange of the mangoes, the fish gave Mama Coty the pearls that they had taken from the manatee's crown. Finally, the manatee let the ocelots ride on its back across the swamps as fast as possible to the border of the jungle where Mama Coty hoped to rescue Clarita, her daughter.

Fig. 14

III

Once the ocelots were in the jungle, they asked a herd of brocket deer if they had seen a troop of squirrel monkeys.

- Indeed, we saw their darks shapes on the canvas of the trees above us. Indeed, we heard their screaming and shouting as they noisily proceeded on their journey. Indeed, they were carrying a prisoner with them. - The brocket deer replied.

- Do you know where they were going? - Mama Coty asked.

- Indeed, they were heading to meet the human poachers. Humans have taken their Prince and the monkeys give them every young cute creature they kidnap, hoping to rescue their Prince. That is the sad truth, indeed. - Was the deer's answer.

- The squirrel monkeys stole my daughter. Please let us ride on your backs so that we can get to their meeting place before my daughter is delivered to the human poachers. - begged Mama Coty.

- We will gladly do as you request. Nevertheless, we have something to ask in return. Five anteaters have been gathering quartz crystals. They each have the same amount. We want to own the crystals that any of them would give us. - The deer explained.

- How many crystals are those?

- In all, they have 50 crystals. That number includes 10 that they will bring to their King. The rest, they have divided equally among the five of them.

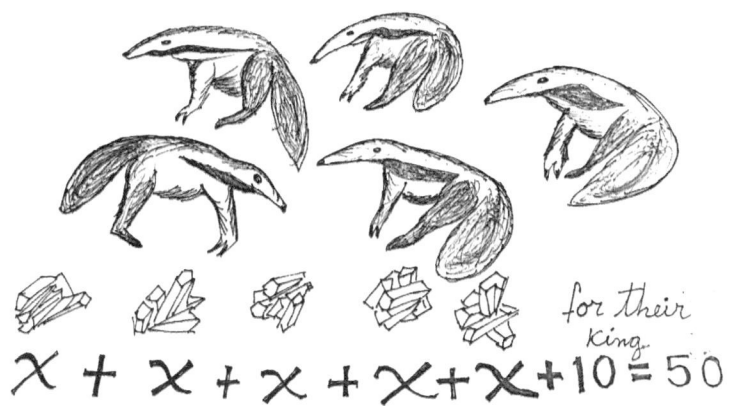

$$X + X + X + X + X + 10 = 50$$

for their king.

5 anteaters with X crystals each +	10 crystals for their king	total
5X +	10 =	50

Fig. 15

Mama Coty looked for the anteaters.

- Will any among you please give me your quartz crystals? The squirrel monkeys stole my daughter. In exchange for the crystals, four brocket deer will let us ride on their backs so that we can get to the squirrel monkeys' meeting place before my daughter is delivered to the human poachers.

- I will gladly give you my quartz crystals, - One of the anteaters said. - but I will ask you for something in return. The river otter collects beautiful shells. She has three boxes, two of them full with the same amount of shells. The third box has not been completed yet, it still lacks six shells to be full. In all, the river otter has 30 shells. I will exchange my quartz crystals for a box full of shells.

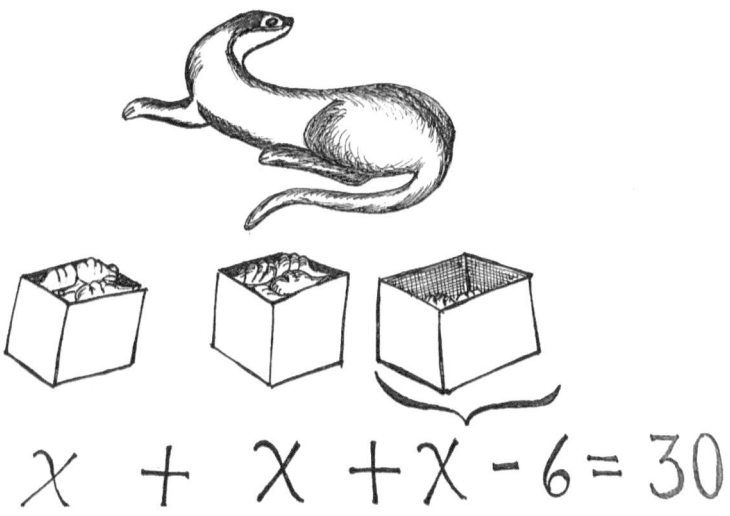

$$X + X + X - 6 = 30$$

3 boxes with   6 are still needed
X shells in each box -   to complete a box   total

3X   -   6   = 30

Fig. 16

Mama Coty went to look for the river otter.

- Will you please give me a box full of shells? An anteater will accept the shells in exchange for its quartz crystals. In exchange for the crystals, four brocket deer will let us ride on their backs so that we can get to the squirrel monkeys' meeting place. The squirrel monkeys stole my daughter. I hope to arrive on time before my daughter is delivered to the human poachers.

- I will gladly give you the shells but I must ask for something in return. The forest snake is the guardian of an ancient Spanish treasure. There are five chests with gold coins and there are other five coins in a small velvet bag. In all, there are 65 coins. If you bring to me the coins of one of the chests, I will gladly give you the shells that you require.

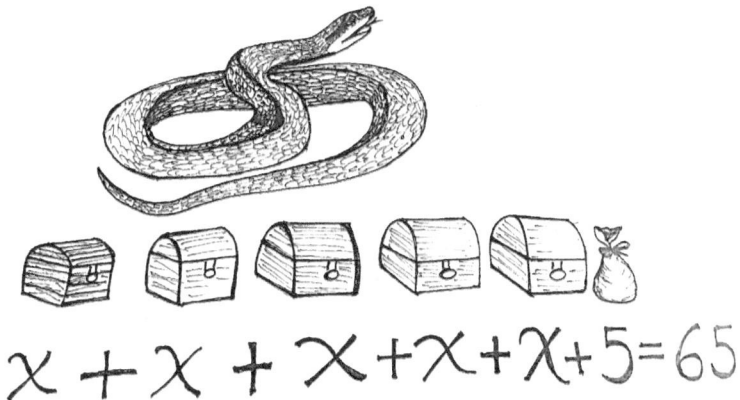

$$x + x + x + x + x + 5 = 65$$

5 chests of coins + one bag of coins = total

    5X   +   5   = 65

Fig. 17

Mama Coty went to look for the snake.

- Will you please give me a chest of gold coins? The river otter asked me for those coins in exchange for a box of shells. An anteater will accept the shells in exchange for its quartz crystals. In exchange for the crystals, four brocket deer will let us ride on their backs so that we can get to the squirrel monkeys' meeting place. The squirrel monkeys stole my daughter. I hope to arrive on time before my daughter is delivered to the human poachers.

- I will gladly give you the coins but I must ask for something in return. The opossum collects sapphires. He has ten bags of the blue stones. Nine bags have the same amount of sapphires but one bag is incomplete. Five sapphires would be needed to complete it. In all, the opossum has 115 sapphires. If you bring to me the sapphires of one of the full bags, I will gladly give you the coins that you require.

$$X + X + X + X + X + X + X + X + X + X - 5 = 115$$

10 bags with sapphires	+	5 needed to complete a bag	=	total
10X	-	5	=	115

Fig. 18

Mama Coty went to look for the opossum.

- Will you please give me a bag full of sapphires? In exchange for them the snake will give me a chest full of gold coins. The river otter asked me for those coins in exchange for a box of shells. An anteater will accept the shells in exchange for its quartz crystals. In exchange for the crystals, four brocket deer will let us ride on their backs so that we can get to the squirrel monkeys' meeting place. The squirrel monkeys stole my daughter. I hope to arrive on time before my daughter is delivered to the human poachers.

- I will gladly give you the sapphires but I must ask for something in return. There is a coati in the jungle. He has three nets full of tangerines and seven tangerines in a tray. In all, the coati has 31 tangerines. If you bring to me the tangerines contained in one of the nets, I will gladly give you the sapphires that you require.

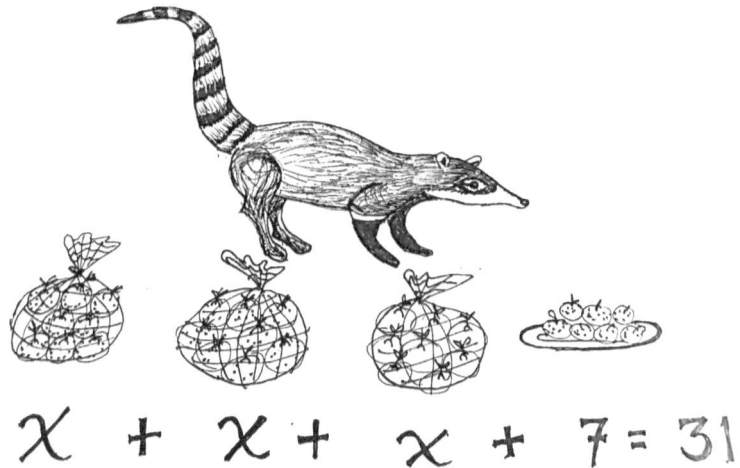

$$X + X + x + 7 = 31$$

3 nets with X tangerines each	+	7 tangerines in a tray	= total
3X	+	7	= 31

Fig. 19

Mama Coty went to look for the coati.

- Will you please give me a net full with tangerines? The opossum wants them to trade for a bag full of sapphires. In exchange for them the snake will give me a chest full of gold coins. The river otter asked me for those coins in exchange for a box of shells. An anteater will accept the shells in exchange for its quartz crystals. In exchange for the crystals, four brocket deer will let us ride on their backs so that we can get to the squirrel monkeys' meeting place. The squirrel monkeys stole my daughter. I hope to arrive on time before my daughter is delivered to the human poachers.

- I will gladly give you the tangerines but I must ask for something in return. The wild turkey stole my wife and he does not want to give her back to me. If you bring back my wife to me, I will gladly give you the sapphires that you require.

Fig. 20

Mama Coty went to look for the wild turkey.

- Will you please let the wife of the coati go free? If you let her free, the coati will give me a net full with tangerines. The opossum wants them to trade for a bag full of sapphires. In exchange for them the snake will give me a chest full of gold coins. The river otter wants those coins in exchange for a box of shells. An anteater will accept the shells in exchange for its quartz crystals. In exchange for the crystals, four brocket deer will let us ride on their backs so that we can get to the squirrel monkeys' meeting place. The squirrel monkeys stole my daughter. I hope to arrive on time before my daughter is delivered to the human poachers.

- I will never let the wife of the coati go free. - Replied the wild turkey.

- I am an ocelot. Ocelots eat wild turkeys. If you do not release the coati's wife, my family and I will eat you for lunch. - Mama Coty warned the turkey.

Fig. 21

Thus, the wild turkey released the coati's wife. The coati was so happy to see his wife again that he gladly gave Mama Coty a net full with tangerines. Once he had the tangerines, the opossum gave the ocelots a bag full of sapphires. In exchange for them, the snake gave Mama Coty a chest full of gold coins. The river otter took the coins and gave the ocelots a box of shells. An anteater received the shells in exchange for its quartz crystals. In exchange for the crystals, four brocket deer let the four ocelots ride on their backs to the squirrel monkeys' palace.

Fig. 22

IV

The troop of squirrel monkeys was already in the Capital City of their Kingdom. The ocelots arrived there on the backs of the brocket deer at the same time as the poachers got there. The poachers and the ocelots entered the royal palace simultaneously. Mama Coty was the first to speak.

- Honorable King of the Squirrel Monkeys, your soldiers have taken my daughter to offer her to these poachers who have also come to meet you.

- You are right. - The King replied - This is a rather unfortunate situation. I will do anything to recover my son, who is kept by the humans.

- There might be a solution for this conundrum. - The chief of the poachers addressed the King. - Noble King, your son is kept as a pet of a young boy, the son of a wealthy businessman. The boy spends too much time playing with the monkey and neglects his studies. His father is very angry with your son for this reason. I am sure that the businessman will gladly give you back your son, the Prince, if you help him.

- How can I help him? - Mama Coty and the King asked simultaneously.

- The boy needs some tutoring in chemistry.

Among the ocelot children, Tere and Esperanza were the brightest of the students and excelled in their grades in every subject.

- We will help the human student to learn chemistry so that he releases the Squirrel Monkey Prince but you must give us our sister back. - Tere and Esperanza said.

- I will keep Clarita as leverage until my son is sound and safe back home. - The King pronounced.

Clarita was kept as a hostage of the King of the squirrel monkeys.

Fig. 23

Mama Coty remained in the squirrel monkey city with Felix, her younger brother. Tere and Esperanza travelled with the poachers to the human town to teach chemistry to the businessman's son.

The poachers took the ocelot girls to the businessman's house. The businessman explained the situation to them.

- If my son does not hand in this chemistry homework, he will fail the course and he will not be admitted in high school.

- I need to learn how to write the correct chemical formula of a compound using the octect rule. - The human student explained to the ocelot girls.

Fig. 24

- I have an idea. - The businessman said. - Let's make a contest. Here is my son's homework, a list of exercises. The one among you, Tere and Esperanza, who better explains the exercises to my son, will get a prize. I will give her a full supply of the best milk chocolate with hazelnuts in the world. Do you agree?

And so, the contest began.

V

First it was Esperanza's turn to explain the octect rule to the son of the businessman.

Instructions: There are boxes that must be filled with eight cupcakes each. Two types of participants contribute with their cupcakes to fill the boxes. How many participants of each type must there be to complete the boxes? Here are three examples.

Example 1.

Example 1

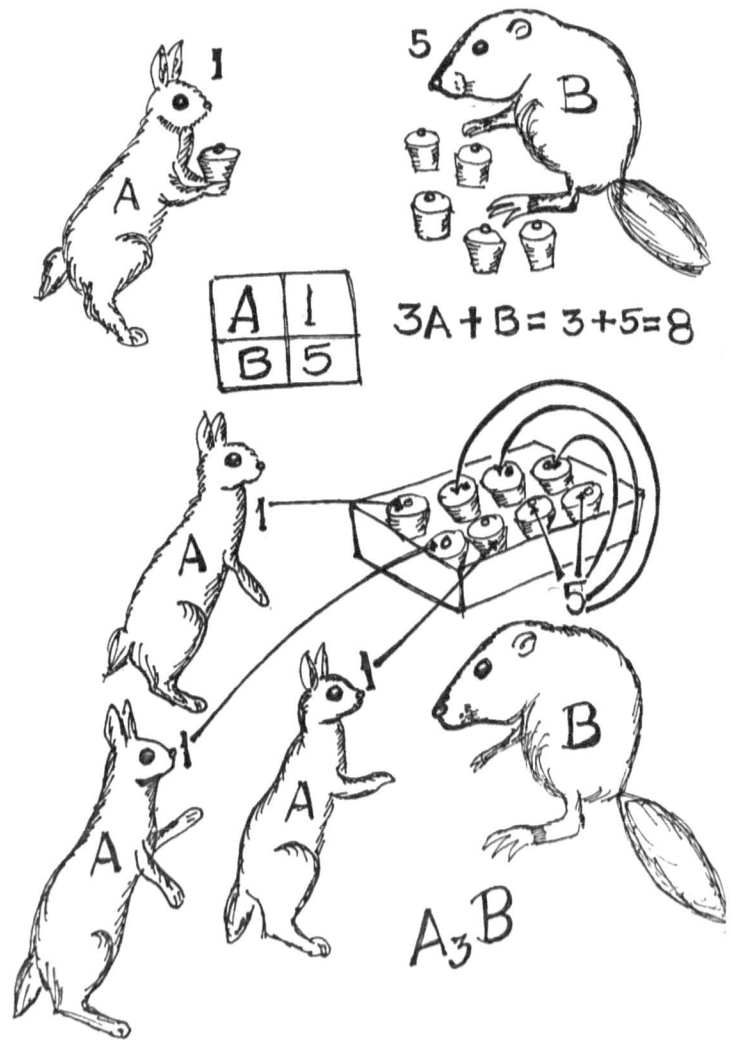

$$3A + B = 3 + 5 = 8$$

Fig. 25.

Example 2.

Each Normal ant (Na) has a cupcake. Each Panda with four Owls on its back (PO_4) has five cupcakes.

How many Na and how many PO_4 are needed to have one complete box of cupcakes?

Example 2

Fig. 26.

In real life, each cupcake is a valence electron. The box is a valence shell. Na is an atom of sodium and PO_4 is a phosphate. Three atoms of sodium and one phosphate are needed to complete the eight valence electrons of sodium phosphate Na_3PO_4.

Example 3.

Each Mongoose (Mn) has three cupcakes. Each Owl (O) has six cupcakes.

How many Mn and how many O are needed to have three complete boxes of cupcakes?

Example 3

Fig. 27.

In real life, each cupcake is a valence electron. The box is a valence shell. Mn is an atom of manganese and O is an atom of oxygen. Two atoms of manganese and three atoms of oxygen are needed to complete the eight valence electrons of manganese (III) oxide Mn_2O_3.

Exercise 1.

Each Horse (H) has one cupcake. Each Owl (O) has six cupcakes.

How many H and how many O are needed to have a complete box of cupcakes?

Exercise 1

Fig. 28.

In real life, each cupcake is a valence electron. The box is a valence shell. H is an atom of hydrogen and O is an atom of oxygen. Two atoms of hydrogen and one atom of oxygen are needed to complete the eight valence electrons of a molecule of water H_2O.

Exercise 2.

Each Fishing elephant (Fe) has three cupcakes. Each Owl (O) has six cupcakes.

How many Fe and how many O are needed to have complete boxes of cupcakes?

Exercise 2

Fig. 29.

In real life, each cupcake is a valence electron. The box is a valence shell. Fe is an atom of iron and O is an atom of oxygen. Two atoms of iron and three atoms of oxygen are needed to complete the eight valence electrons of iron(III) oxide Fe_2O_3.

Exercise 3.

Each Horse (H) has one cupcake. Each Skunk with four owls on its back (SO_4) has six cupcakes.

How many H and how many SO_4 are needed to have complete boxes of cupcakes?

Exercise 3

Fig. 30.

In real life, each cupcake is a valence electron. The box is a valence shell. H is an atom of hydrogen and SO_4 is a sulfate. Two atoms of hydrogen and one sulfate are needed to complete the eight valence electrons of sulfuric acid H_2SO_4.

Exercise 4.

Each Zebra narwhal (Zn) has two cupcakes. Each Skunk with four owls on its back (SO_4) has six cupcakes.

How many Zn and how many SO_4 are needed to have complete boxes of cupcakes?

Exercise 4

Fig. 31.

In real life, each cupcake is a valence electron. The box is a valence shell. Zn is an atom of zinc and SO_4 is a sulfate. One atom of zinc and one sulfate are needed to complete the eight valence electrons of zinc sulfate $ZnSO_4$.

Exercise 5.

Each Canary (Ca) has two cupcakes. Each Camel with three owls on its back (CO_3) has six cupcakes.

How many Ca and how many CO_3 are needed to have complete boxes of cupcakes?

Exercise 5

Fig. 32.

In real life, each cupcake is a valence electron. The box is a valence shell. Ca is an atom of calcium and CO_3 is a carbonate. One atom of calcium and one carbonate are needed to complete the eight valence electrons of calcium carbonate $CaCO_3$.

Exercise 6.

Each Koala (K) has one cupcake. Each Skunk with four owls on its back (SO_4) has six cupcakes.

How many K and how many SO_4 are needed to have complete boxes of cupcakes?

Exercise 6

Fig. 33.

In real life, each cupcake is a valence electron. The box is a valence shell. K is an atom of potassium and SO_4 is a sulfate. Two atoms of potassium and one sulfate are needed to complete the eight valence electrons of potassium sulfate K_2SO_4.

Exercise 7.

Each Astonishing goat (Ag) has one cupcake. Each Nightingale married to three owls (NO_3) has seven cupcakes.

How many Ag and how many NO_3 are needed to have complete boxes of cupcakes?

Exercise 7

Fig. 34.

In real life, each cupcake is a valence electron. The box is a valence shell. Ag is an atom of silver and NO_3 is a nitrate. One atom of silver and one nitrate are needed to complete the eight valence electrons of silver nitrate $AgNO_3$.

Exercise 8.

Each Normal ant (Na) has one cupcake. Each Nightingale married to three owls (NO_3) has seven cupcakes.

How many Na and how many NO_3 are needed to have complete boxes of cupcakes?

Exercise 8

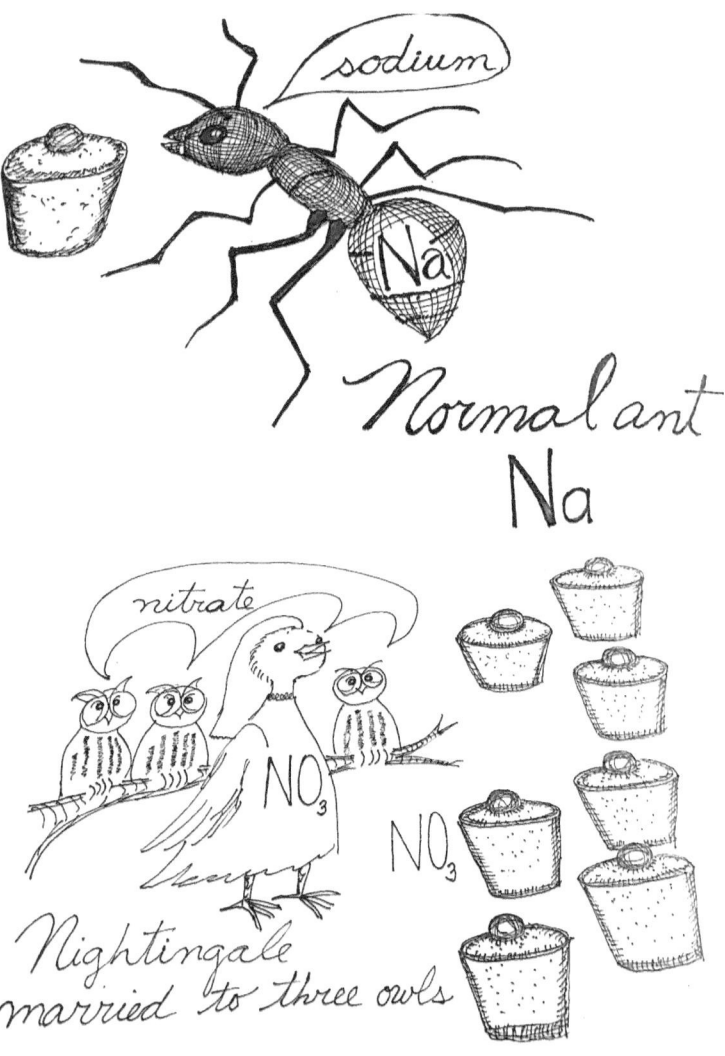

Fig. 35.

In real life, each cupcake is a valence electron. The box is a valence shell. Na is an atom of sodium and NO_3 is a nitrate. One atom of sodium and one nitrate are needed to complete the eight valence electrons of sodium nitrate $NaNO_3$.

Exercise 9.

Each Horse (H) has one cupcake. Each Bison with three owls on its back (BO_3) has five cupcakes.

How many H and how many BO_3 are needed to have complete boxes of cupcakes?

Exercise 9

Fig. 36.

In real life, each cupcake is a valence electron. The box is a valence shell. H is an atom of hydrogen and BO_3 is a borate. Three atoms of hydrogen and one borate are needed to complete the eight valence electrons of boric acid H_3BO_3.

Exercise 10.

Each Koala (K) has one cupcake. Each Cute lamb (Cl) with three Owls on its back (ClO_3) has seven cupcakes.

How many K and how many ClO_3 are needed to have complete boxes of cupcakes?

Exercise 10

Fig. 37.

In real life, each cupcake is a valence electron. The box is a valence shell. K is an atom of potassium and ClO_3 is a chlorate. One atom of potassium and one chlorate are needed to complete the eight valence electrons of potassium chlorate $KClO_3$.

Exercise 11.

Each Normal ant (Na) has one cupcake. Each Mongoose with four owls on its back (MnO_4) has six cupcakes.

How many Na and how many MnO_4 are needed to have complete boxes of cupcakes?

Exercise 11

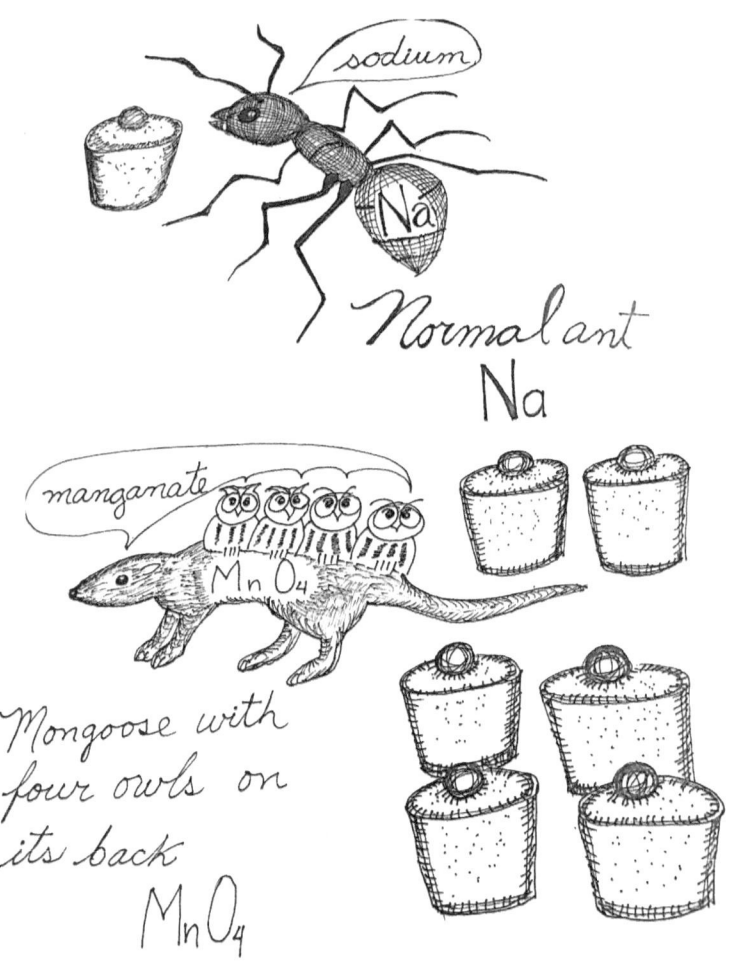

Fig. 38.

In real life, each cupcake is a valence electron. The box is a valence shell. Na is an atom of sodium and MnO_4 is a manganate. Two atoms of sodium and one manganate are needed to complete the eight valence electrons of sodium manganate Na_2MnO_4.

Exercise 12.

Each Magpie (Mg) has two cupcakes. Each Owl married to a Horse (OH) has seven cupcakes.

How many Mg and how many OH are needed to have complete boxes of cupcakes?

Exercise 12

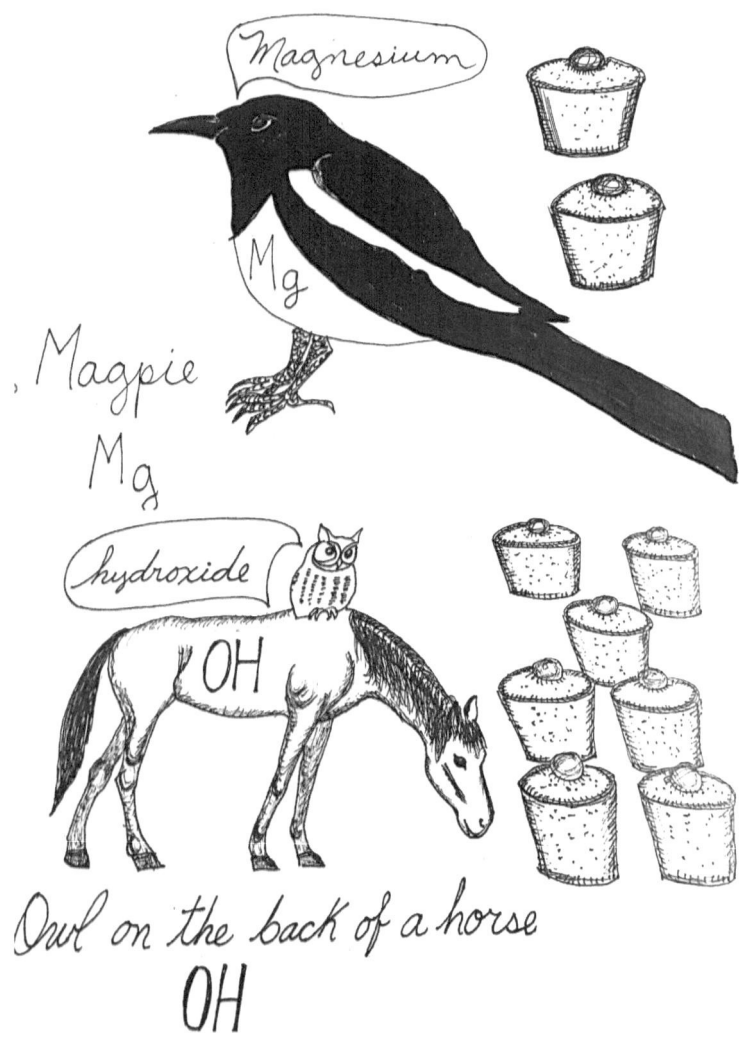

Fig. 39.

In real life, each cupcake is a valence electron. The box is a valence shell. Mg is an atom of magnesium and OH is a hydroxide. One atom of magnesium and two hydroxides are needed to complete the eight valence electrons of magnesium hydroxide $Mg(OH)_2$.

VI

Then it was Tere's turn to explain the chemical formulas to the son of the businessman.

Instructions: There are two different types of characters. One type has a debt of coins. The other type has a certain amount of coins. How many of the second type are needed to pay the debt? Here are three examples.

Example 1.

Example 1

Type A has a debt of certain amount of coins. Type B has a certain amount of coins.

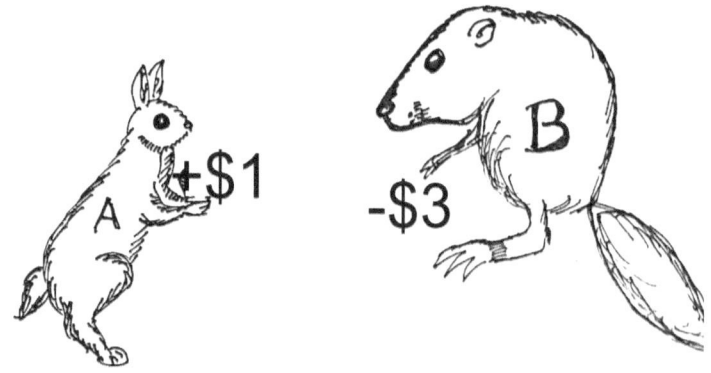

How many individuals of type B are needed to pay type A's debt?

Fig. 40.

Example 2.

Each Normal ant (Na) has a coin. Each Panda with four Owls on its back (PO_4) has to pay three coins.

How many Na are needed so that one PO_4 can pay his debt?

Example 2

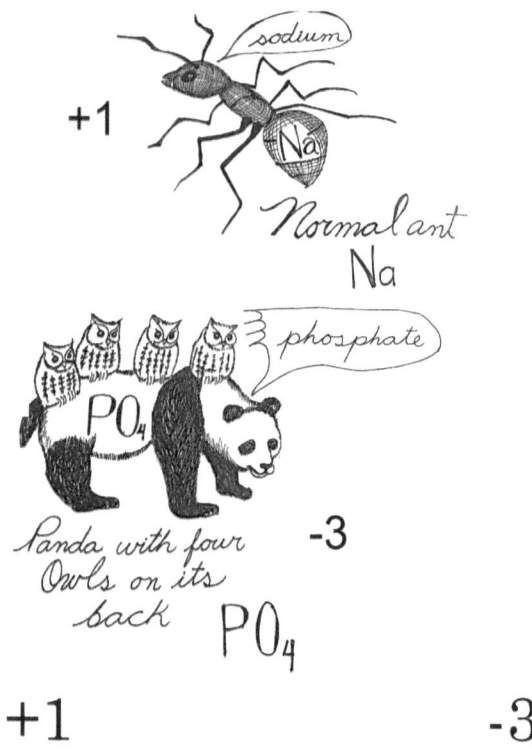

+1

Normal ant
Na

Panda with four
Owls on its
back   -3
PO_4

+1   -3

Total debt = -3
How many Na are needed so that one
PO_4 can pay his debt?
Three Na are needed so that one PO_4
can pay his debt.

Fig. 41.

In real life, Na is an atom of sodium and PO_4 is a phosphate. Three atoms of sodium and one phosphate are needed to have a molecule of sodium phosphate Na_3PO_4.

Example 3.

Each Mongoose (Mn) has three coins. Three Owls (O) have to pay two coins each.

How many Mn are needed so that the three O can pay their debt?

Example 3

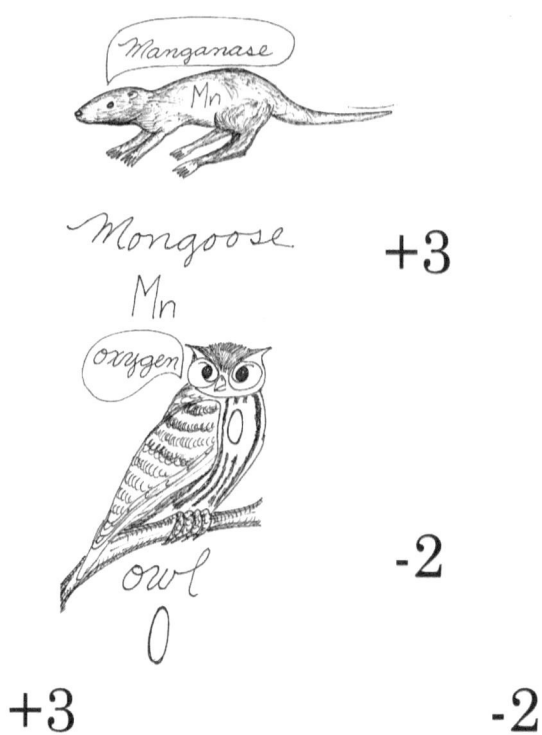

+3

+3   -2

Total debt = - 6
How many Mn are needed so that the
three O can pay their debt?
Two Mn are needed so that three O
can pay their debt.

Fig. 42.

In real life, Mn is an atom of manganese and O is an atom of oxygen. Two atoms of manganese and three atoms of oxygen are needed to have a molecule of manganese (III) oxide Mn_2O_3.

Exercise 1.

Each Horse (H) has one coin. Each Owl (O) has to pay two coins.

How many H are needed so that one O can pay his debt?

Exercise 1

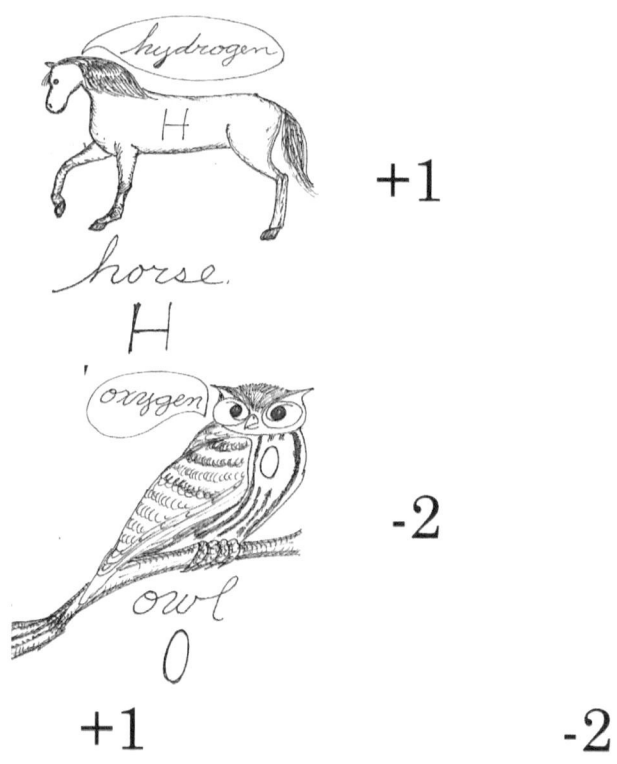

+1   -2

Total debt = -2

How many H are needed so that one O can pay his debt?

Two H are needed so that one O can pay his debt.

Fig. 43.

In real life, H is an atom of hydrogen and O is an atom of oxygen. Two atoms of hydrogen and one atom of oxygen are needed to have a molecule of water H_2O.

Exercise 2.

Each Fishing elephant (Fe) has three coins. Three Owls (O) have to pay two coins each.

How many Fe are needed so that the three O can pay their debt?

Exercise 2

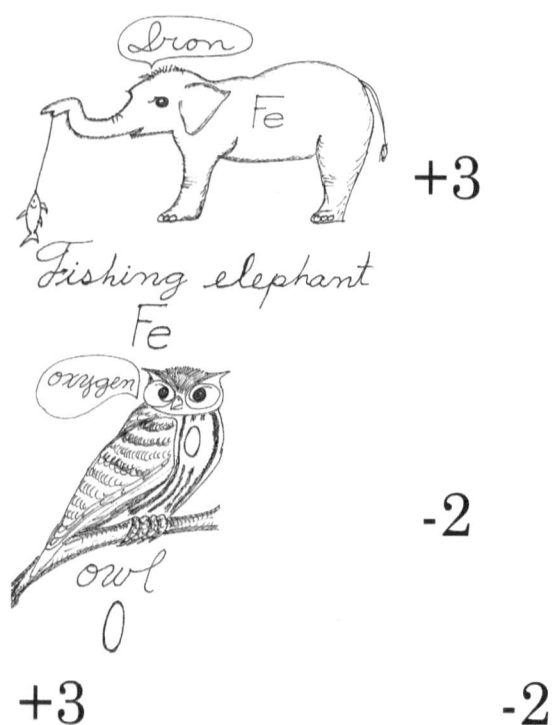

+3

-2

+3   -2

Total debt = -6
How many Fe are needed so that
three O can pay their debt?
Two Fe are needed so that three O can
pay their debt.

Fig. 44.

In real life, Fe is an atom of iron and O is an atom of oxygen. Two atoms of iron and three atoms of oxygen are needed to have a molecule of iron(III) oxide Fe_2O_3.

Exercise 3.

Each Horse (H) has one coin. Each Skunk with four owls on its back (SO_4) has to pay two coins.

How many H are needed so that one SO_4 can pay his debt?

Exercise 3

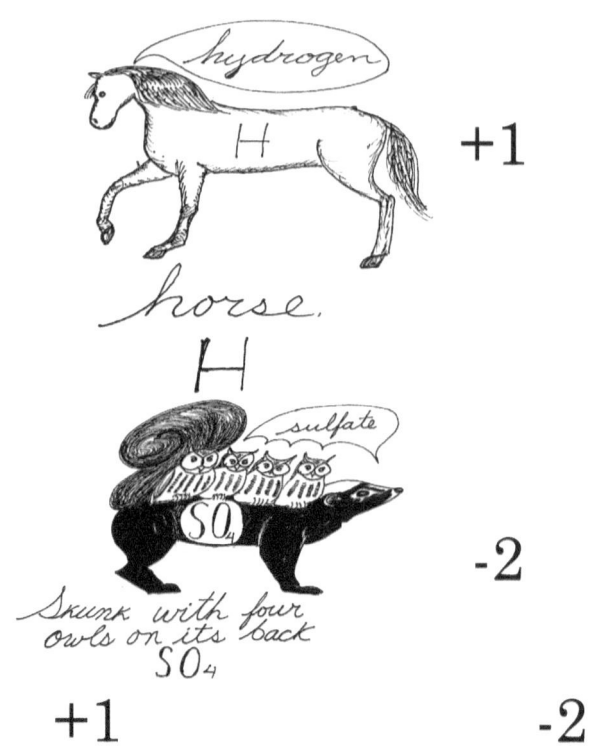

+1

-2

+1   -2

Total debt = -2
How many H are needed so that one
SO_4 can pay his debt?
Two H are needed so that one SO_4 can
pay his debt.

Fig. 45.

In real life, H is an atom of hydrogen and SO_4 is a sulfate. Two atoms of hydrogen and one sulfate are needed to have a molecule of sulfuric acid H_2SO_4.

Exercise 4.

Each Zebra narwhal (Zn) has two coins. Each Skunk with four owls on its back (SO_4) has to pay two coins.

How many Zn are needed so that one SO_4 can pay his debt?

Exercise 4

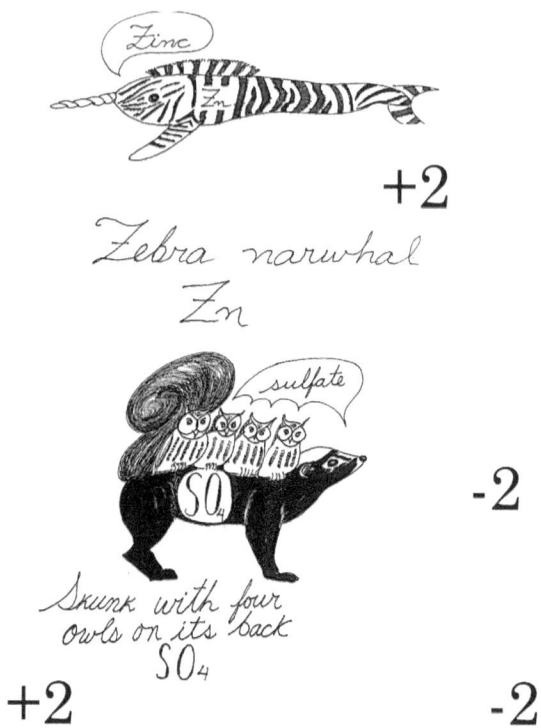

$+2$

Zebra narwhal
Zn

-2

Skunk with four
owls on its back
SO_4

$+2$ -2

Total debt = -2
How many Zn are needed so that one
SO_4 can pay his debt?
One Zn is needed so that one SO_4 can
pay his debt.

Fig. 46.

In real life, Zn is an atom of zinc and SO_4 is a sulfate. One atom of zinc and one sulfate are needed to have a molecule of zinc sulfate $ZnSO_4$.

Exercise 5.

Each Canary (Ca) has two coins. Each Camel with three owls on its back (CO_3) has to pay two coins.

How many Ca are needed so that one CO_3 can pay his debt?

Exercise 5

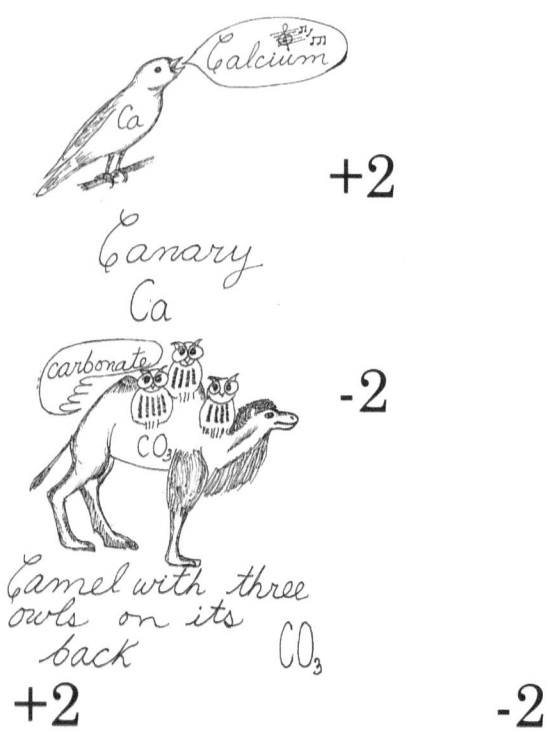

+2

-2

+2   -2

Total debt = -2
How many Ca are needed so that one
CO_3 can pay his debt?
One Ca is needed so that one CO_3 can
pay his debt.

Fig. 47.

In real life, Ca is an atom of calcium and CO_3 is a carbonate. One atom of calcium and one carbonate are needed to have a molecule of calcium carbonate $CaCO_3$.

Exercise 6.

Each Koala (K) has one coin. Each Skunk with four owls on its back (SO_4) has to pay two coins.

How many K are needed so that one SO_4 can pay his debt?

Exercise 6

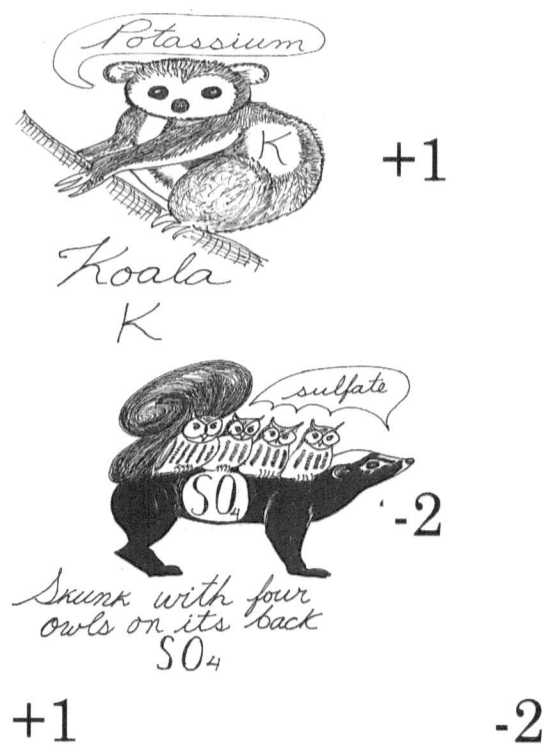

+1

-2

+1   **-2**

Total debt = -2
How many K are needed so that one
SO_4 can pay his debt?
Two K are needed so that one SO_4 can
pay his debt.

Fig. 48.

In real life, K is an atom of potassium and SO_4 is a sulfate. Two atoms of potassium and one sulfate are needed to have a molecule of potassium sulfate K_2SO_4.

Exercise 7.

Each Astonishing goat (Ag) has one coin. Each Nightingale married to three owls (NO_3) has to pay one coin.

How many Ag are needed so that one NO_3 can pay his debt?

Exercise 7

+1

Astonishing goat

Ag

-1

Nightingale married to three owls

+1   **-1**

Total debt = -1
How many Ag are needed so that one
NO_3 can pay his debt?
One Ag is needed so that one NO_3 can
pay his debt.

Fig. 49.

In real life, Ag is an atom of silver and NO_3 is a nitrate. One atom of silver and one nitrate are needed to have a molecule of silver nitrate $AgNO_3$.

Exercise 8.

Each Normal ant (Na) has one coin. Each Nightingale married to three owls (NO_3) has to pay one coin.

How many Na are needed so that one NO_3 can pay his debt?

Exercise 8

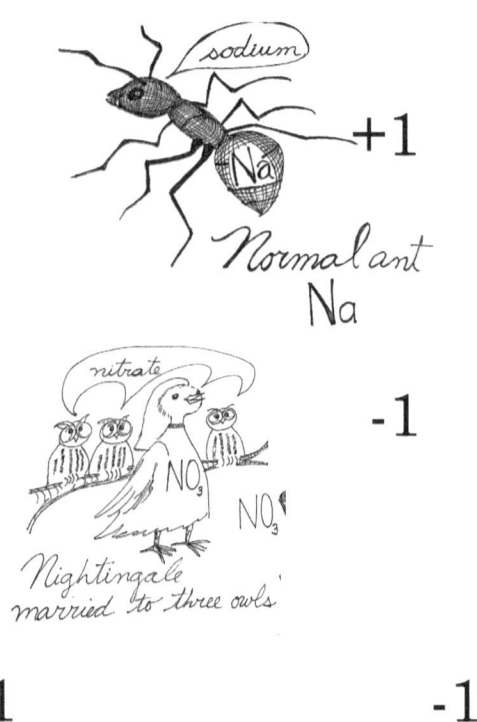

sodium

+1

Normal ant
Na

nitrate

-1

Nightingale
married to three owls

+1   -1

Total debt = -1
How many Na are needed so that one
NO_3 can pay his debt?
One Na is needed so that one NO_3 can
pay his debt.

Fig. 50.

In real life, Na is an atom of sodium and NO_3 is a nitrate. One atom of sodium and one nitrate are needed to have a molecule of sodium nitrate $NaNO_3$.

Exercise 9.

Each Horse (H) has one coin. Each Bison with three owls on its back (BO_3) has to pay three coins.

How many H are needed so that one BO_3 can pay his debt?

Exercise 9

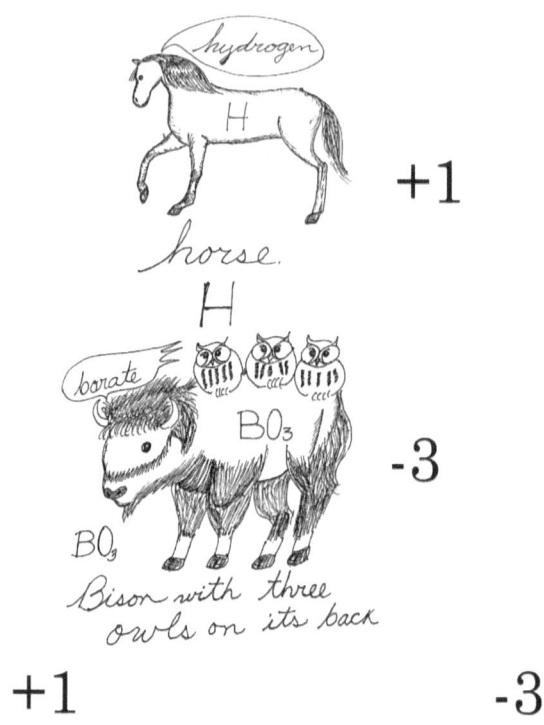

+1

−3

+1   −3

Total debt = -3
How many H are needed so that one
BO$_3$ can pay his debt?
Three H are needed so that one BO$_3$
can pay his debt.

Fig. 51.

In real life, H is an atom of hydrogen and BO_3 is a borate. Three atoms of hydrogen and one borate are needed to have a molecule of boric acid H_3BO_3.

Exercise 10.

Each Koala (K) has one coin. Each Cute lamb (Cl) with three Owls on its back (ClO_3) has to pay one coin.

How many K are needed so that one ClO_3 can pay his debt?

Exercise 10

Total debt = -1
How many K are needed so that one
ClO$_3$ can pay his debt?
One K is needed so that one ClO$_3$ can
pay his debt.

Fig. 52.

In real life, K is an atom of potassium and ClO_3 is a chlorate. One atom of potassium and one chlorate are needed to have a molecule of potassium chlorate $KClO_3$.

Exercise 11.

Each Normal ant (Na) has one coin. Each Mongoose with four owls on its back (MnO_4) has to pay two coins.

How many Na are needed so that one MnO_4 can pay his debt?

Exercise 11

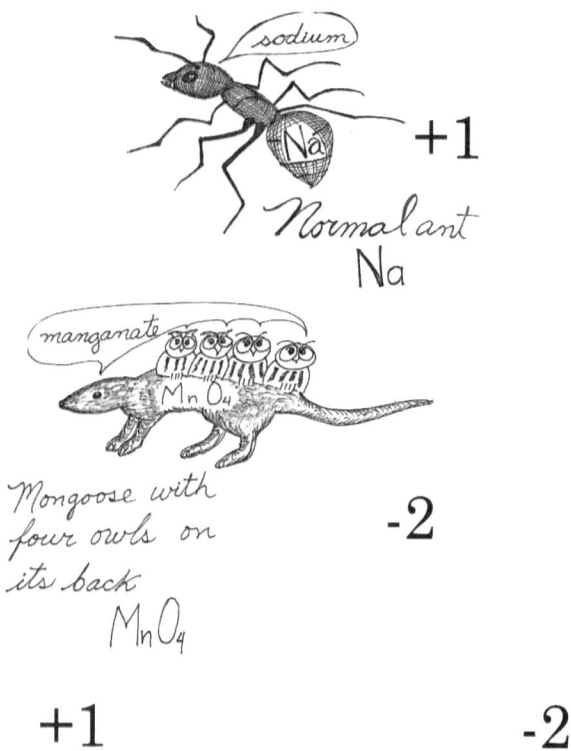

+1

Na

-2

+1   -2

Total debt = -2
How many Na are needed so that one
MnO_4 can pay his debt?
Two Na are needed so that one MnO_4
can pay his debt.

Fig. 53.

In real life, Na is an atom of sodium and MnO_4 is a manganate. Two atoms of sodium and one manganate are needed to have a molecule of sodium manganate Na_2MnO_4.

Exercise 12.

Each Magpie (Mg) has two coins. Two Owls, each married to a Horse (OH), have to pay one coin each.

How many Mg are needed so that the two OH can pay their debt?

Exercise 12

Total debt = -2
How many Mg are needed so that the
two OH can pay their debt?
One Mg is needed so that the two OH
can pay their debt.

Fig. 54.

In real life, Mg is an atom of magnesium and OH is a hydroxide. One atom of magnesium and two hydroxides are needed to have a molecule of magnesium hydroxide $Mg(OH)_2$.

VII

Answers to exercises corresponding to Figures 9-17.

Figure 9.   X = 7
Figure 10. X = 5
Figure 11. X = 10
Figure 12. X = 20
Figure 13. X = 8
Figure 14. X = 8
Figure 15. X = 12
Figure 16. X = 12
Figure 17. X = 8